S0-ASH-079

FINLAND

MAJOR WORLD NATIONS
FINLAND

Alan James

CHELSEA HOUSE PUBLISHERS
Philadelphia

914.987
JAM

Chelsea House Publishers

Copyright © 2000 by Chelsea House Publishers,
a division of Main Line Book Co.
All rights reserved.
Printed in Malaysia.

First Printing.

1 3 5 7 9 8 6 4 2

Library of Congress Cataloging-in-Publication Data

James, Alan, 1943-
Finland / Alan James.
p.   cm. — (Major world nations)
Includes index.
Summary: An overview of the history, geography, economy, government,
people, and culture of Finland.
ISBN 0-7910-5384-9
1. Finland—Juvenile literature.   [1. Finland.]   I. Title.
II. Series.
DL1012.J35   1999
948.97—dc21   99-19152
CIP

ACKNOWLEDGEMENTS

The Author and Publishers are grateful to the following organizations and individuals
for permission to reproduce copyright illustrations in this book:
Joan Todd; Tom Söderman, Press Councillor at the Embassy of Finland in London.

CONTENTS

FACTS AT A GLANCE

Land and People

Official Name   Republic of Finland

Location   Northernmost country on Scandinavian peninsula; one-third of country is above Arctic Circle

Area   130,000 square miles

Climate   Subarctic with very short, warm summers

Capital   Helsinki

Other Cities   Turku, Tampere

Population   5 million

Population Density   38 persons per square mile

Major Rivers   Kemi, Torne, Muonio, Oulu, Vuoski

Major Lakes   Saimaa, Ladoga, Paijanne (Finland has over 187,888 lakes)

Highest Point   Halti (4,344 feet above sea level)

Official Languages   Finnish and Swedish

Other Languages   English

7

Ethnic Groups	Finns, 91 percent; Swedes, 8 percent
Religions	Evangelical Lutheran, 89 percent
Literacy Rate	Nearly 100 percent

Economy

Natural Resources	Timber, copper, zinc, iron ore, silver
Division of Labor Force	Public services, 30.4 percent; industry 20.9 percent; commerce, 15 percent; business/finance services, 10.2 percent; agriculture, 8.6 percent; transportation/communication 7.7 percent; construction, 7.2 percent
Agricultural Products	Grains, sugar beets, potatoes, dairy cattle, fish
Industries	Metal products, paper, shipbuilding, copper refining, food processing, textiles, chemicals
Major Imports	Food, petroleum, chemicals, transportation equipment, iron and steel, machinery
Major Exports	Paper products, chemicals, timber, metals, high-tech industrial products
Major Trading Partners	European Union, Sweden, United States, Japan, Russia
Currency	Markka

Government

Form of Government	Republic
Formal Head of State	President
Head of Government	Prime minister
Voting Right	All citizens 18 years of age or older

8

HISTORY AT A GLANCE

1st century B.C.   Farmers begin to clear some of the forest land for growing crops in the area that one day would become Finland.

800-1100 A.D.   The Vikings arrive in the area and establish trade routes. Many different cultures come to Finland and there are many disputes between the Finnish tribes that gradually form.

1155   King Eric IX of Sweden and the Bishop of Uppsala take over the land of the Finns and begin to convert the people to Christianity. Finland becomes part of the kingdom of Sweden.

14th-16th centuries   Under Swedish rule Finland prospers. Helsinki becomes part of one of the most important Swedish trade routes. The religious beliefs of Lutheranism are introduced by Michael Agricola.

1581   Finland is given the status of a Grand Duchy by the Swedish king, John III.

1596-97   The peasants revolt because of quarrels over social conditions, foreign policy, and religious differences. Called the Club War, the peasants are crushed.

1676	A great famine kills nearly one-third of the population of Finland.
18th century	Russia and Sweden wage numerous wars, often making Finland their battleground.
1808	Russia seizes all of Finland from Swedish control and it becomes a Grand Duchy of Russia. Russia allows Finland to rule itself without much interference.
1906	Finland gives women the right to vote. Its parliament is changed to a one-chamber body.
1917	Taking advantage of the Russian Revolution the Finns call a general strike which leads to an internal war. On December 6 the Finnish issue their Declaration of Independence from Russia.
1918	The war and independence leave Finland in dire straits economically and there are widespread food storages.
1919	Finland adopts a new constitution making it a republic.
1920	Finland joins the League of Nations, a precursor of the United Nations. A peace-treaty is signed with the Soviet Union.
1939-1940	The 14-week Winter War between the Soviet Union and Finland is fought at the start of World War II. Though the Finns fight bravely the Russian invaders win. Afterward the Treaty of Moscow gives the Soviet Union a large part of southeastern Finland.
1941-44	Finland continues to battle the Soviet Union throughout World War II sometimes with the

help of Germany. Finland allows German soldiers to use its land during the war. Carl Mannerheim is the military leader at this time.

1944   As the Germans are forced out of Finland they devastate the area of Lapland, destroying everything in their path. After the war Finland is forced to give up the territory of Karelia and the port of Petsamo to the Soviet Union. Mannerheim becomes president of Finland.

1947   A peace treaty is signed with the Soviet Union. Finland gives the Russians 12 percent of its land. Many of the Finns in the surrendered territory move with all their belongings into Finland rather than live under Soviet rule.

1950s   Finland gradually changes from an agricultural country into an industrial one.

1952   The city of Helsinki hosts the Olympic Games. Finland announces a policy of neutrality.

1955   Finland joins the United Nations.

1956-1982   Urho Kekkonen is president of Finland during the difficult Cold War era. Finland continues a policy of neutrality, diplomatically balancing relations between East and West.

1962   The five Scandinavian countries, Finland, Sweden, Denmark, Norway, and Iceland, sign an agreement to strengthen cooperation between them.

1975   A conference on security and cooperation is held in Helsinki with 35 countries attending. The Helsinki Accord is signed.

1982   President Kekkonen resigns due to ill health and Mauno Koivisto succeeds him as president.

1988   Koivisto is reelected for a second six-year term.

1992   Finland celebrates its 75th year of independence. Russia and Finland sign an agreement that pledges to settle disputes between them peacefully.

1995   Finland joins the European Union with most other European nations.

1

Introducing Finland

Finland is one of the most northerly countries in the world lying between the 60th and 70th degrees of latitude. About one-third of the total length of the country lies north of the Arctic Circle. To the west is Sweden, to the north is the tip of Norway, and to the east is Russia.

To the south and southwest is the Baltic Sea with its two arms, the Gulf of Finland and the Gulf of Bothnia. The Baltic Sea is only a sea in summer since it is mostly frozen solid in winter. In a severe winter it is possible to drive a car across the thick ice for more than 50 miles (80 kilometers) from Finland toward Sweden. The Baltic has many rivers pouring into it and so much fresh water reduces the salt content and causes the sea to freeze very quickly in winter. The Finnish ports of Turku, Helsinki, and Hanko all have channels kept open to shipping during the winter by a fleet of ice-breakers on loan from other ports which are regularly icebound in the depths of winter.

Finland is larger than the British Isles, the total area comprising

130,000 square miles (337,000 square kilometers); all of this for a population of only 5 million. Finland (or *Suomi* in Finnish) is a relatively low-lying country, and is mostly covered with dense forests. The soil is poor, mainly consisting of deposits left after the Ice Age. In many places the soil is so thin that the surface contours of the land often follow the layers of bedrock. The land rises slightly in Lapland in the north where there are hills called fells.

There are about 30,000 Finnish islands mainly off the south and southwest coasts. Finland is the land of lakes and forests.

A Baltic icebreaker leading a convoy of ships through the thick ice in winter. It is really only a sea in summer.

A Lapp riding a snow scooter.

There are about 60,000 lakes which account for 9 percent of the total area of the country. There are so many lakes that those too small to be given a name are not even included in this number. The lakes were carved from the granite bedrock when the glaciers retreated at the end of the Ice Age. There are lakes of all sizes joined by short rivers or rapids, channels or canal, and they form a continuous waterway on which boats can travel for great distances. The large numbers of lakes make traveling and the planning of roads and railways a difficulty. Roads frequently come to an abrupt end and vehicles must continue their journey by ferry; and this is especially so in the southeast of Finland where the lakes are at their most plentiful, in places occupying almost half the surface area of the land.

15

A small part of Lake Inari in Finnish Lapland—a lake so large that it has more than 3,000 islands in it.

There are three main groups of lakes. The largest is in the east where Lake Saimaa and thousands of smaller lakes feed the Vuoksi River and Lake Ladoga in Russia. Lake Päijänne is the main lake in the central system. From these lakes the water reaches south to the Gulf of Finland down the system of rapids on the Kymi River. The group of lakes in the west, with Pyhäjärvi as its largest, drains to the Gulf of Bothnia through the Kokemäenjoki (*joki* means *river* in Finnish).

The nearness of the Atlantic Ocean, the North Atlantic Drift, and warm air currents explain why Finland has average temperatures which are much higher than in other parts of the world at the same latitude. Finland's climate is one of warm summers and

16

cold winters. At midsummer, southern Finland has as much as 19 hours of daylight. The warmest month is July (maximum 30 degrees Celsius) and the coldest is February (minimum *minus* 30 degrees Celsius).

In the north of Scotland the sun rises at three o'clock in the morning and sets at 10 at night on the longest day of the year. So the sun falls below the horizon for only about five hours on that day. But in even more northerly places, including Lapland, the sun does not set at all from the second week of May to the last week in July, resulting in daylight—both during the day and at night—for as much as 10 weeks in the year at Utsjoki in the far north of Lapland. At that time the countryside is ablaze with the golden fire of the Midnight Sun.

This constant daylight is paid for in the far north by continuous winter darkness. Between the end of November and the end of January, the sun does not rise at all. During this period of the year

The midnight sun.

Logs are brought to the river by lorry and then tipped into the water. Finland is a land of forests and the timber industry is very important.

electric lighting is the only means by which normal life can be carried on. Some industries in the far north are kept going in winter by the use of floodlighting. Summer lasts from June until August, and after autumn has ended the whole country is snowbound for four months every year. (Lapland in the north may be snowbound for seven months.) During the depths of winter (December to March) the temperature usually stays permanently below freezing point, even on the milder coast of the southwest.

The average annual precipitation, the amount of snow and rain, is 27 inches (700 millimeters) in the south and 16 inches (400 millimeters) in Lapland. Much of Lapland is called tundra—flat and boggy land. Only the top two feet (62 centimeters) of soil unfreezes in the short summer. Below that the ground is frozen solid permanently. This frozen ground is called permafrost. It is difficult to sink the foundations of buildings into earth that is rock-hard and frozen. The tundra is treeless—although smaller bushes do grow there—because large trees would find it impossible to work their roots deep into the frozen soil and because of the very cold winters when the temperature may be below zero for perhaps eight months of the year.

Finland is also a land of forests; almost three-quarters of the total land area is covered with timber. Going northward, the climate causes spruce to disappear first, then pine. Only dwarf, stunted birch trees can be found in the north of Lapland.

The secret of Finland's beauty is that despite cities, factories, and the introduction of more advanced farming methods, it still remains a simple, empty land of forests and lakes.

2

Early Life in Finland

Primitive man inhabited what is now Finland, as he inhabited much of northern Europe, before 6000 B.C. Life in the Stone Age and the Bronze Age was a life of hunting and fishing. They only ceased to be hunters and assumed the life of settled farmers a little more than 2,000 years ago. Finland has more forests and lakes than almost any other country in Europe, and when man began to live a settled life as a farmer the best furs were found in the pine forests of the north. The ancestors of the present-day Finns once lived south of Finland in Estonia but, in time, many settled in Finland bringing cattle, sheep, and seeds to sow in the ground. Some of them continued to hunt wild animals both for their prized furs and for their meat. First, however, these farmers needed to create the fields in which to sow their seeds and the immediate problem was to clear the land of trees. These early Finns set fire to the trees and burn-beat the land. In time, trees grew up again and burn-beating had to begin once more.

The land which was most successfully cleared of trees was used

Pine trees, which grow over much of Finland and provide the country with valuable timber.

as arable land and grain crops were sown there. Other burn-beaten land was used as pasture for herds of animals. For many hundreds of years life remained very simple. There was little trade and the exchange of goods took place by bartering one thing for something else, without the use of money. Up to the 16th century corn was bartered locally or at the coastal meeting places in exchange for weapons, textiles, salt, and ornaments. But by the end of that century man had found another use for the pine forests near the coast. Some were burnt to obtain pitch for tar which was in heavy demand in the 16th and 17th centuries

21

because of the increase of shipping and of the famous voyages of discovery. Tar was needed to strengthen ropes and to seal joints in the sides of wooden ships.

By then, too, wood of various kinds was being exported to other countries. It is strange that the timber industry did not grow really important in Finland until the 19th century, by which time huge sawmills and pulp works had become essential. Today, the forest industry accounts for half the total exports from Finland.

It is often said that Finnish history has been one of oppression since the Middle Ages. As early as 1155, a Swedish crusade to Finland was led by Bishop Henry of Uppsala who was born an Englishman and who is now regarded as St. Henry, the patron saint of Finland. This religious mission in the 12th century led to Finland becoming part of the kingdom of Sweden, and this situation persisted until the early 19th century. In 1809, as a result of war, Sweden surrendered Finland to Russia when the land of the Finns became a

This sophisticated machinery can deal with huge quantities of timber.

A typical home interior of several centuries ago. Notice one man sitting up in bed and another playing a kantele—Finland's national instrument.

grand duchy, the czar of Russia being known as the Grand Duke of Finland. Foreign affairs were controlled by Russia, but within Finland itself the Finns made their own decisions.

For 600 years, while Finland had been part of the kingdom of Sweden, Swedish had been the official language. In 1809 when Finland came under Russian rule the Finns began their slow struggle toward independence. One way of doing this was to encourage the use of their mother tongue. This was summed up by the slogan: "We are no longer Swedes. We cannot become Russians. We must be Finns." Then came more books written in Finnish and, by 1863, the language was officially recognized.

23

3

A Free Land

Against the background of Russian rule, Finland had its own government, law, courts, postal service, army, and currency. In 1906, the old parliament (known as the Diet of the Four Estates—Nobles, Clergy, Burghers, and Peasants) was replaced by a parliament with a single chamber. In the same year women were given the right to vote and to run for office. (Finland was the first country in Europe to grant women the right to run for office. New Zealand was the first country in the world to give women the right to vote, followed by Finland.) But many Finns remained angry at the fact that their country was ruled by Russia, especially in the early 20th century when the czar of Russia began to make things difficult for Finland. Many Finns reacted by quietly resisting change and also by holding meetings to discuss the need for independence.

Finland remained in the hands of the Russians until the Finnish Declaration of Independence on December 6, 1917. (Finland took advantage of the problems inside Russia created by the Bolshevik

Revolution to declare its independence.) But history is never a simple matter. Some Finns joined with the Russian soldiers in the country against the rest of Finland and there was civil war in the spring of 1918–the "whites" eventually defeated the "reds" and so Finland remained a free nation of Scandinavia.

Finland became an independent republic in 1919 when the present political constitution came into force. Peace was signed with Russia in 1920. The Finnish flag is a blue cross on a white background and was adopted in 1918. The next 20 years was a period of social and economic progress as the country found out how to rule itself and make its own decisions.

Finland again found itself in conflict with the Soviet Union after the Second World War (1939-45) had broken out. This resulted in the 14-week "Winter War" in 1939-40, partly caused by the Soviet

Ski troops fighting the Russians in 1940.

demand to use the port of Hanko in the southwest of Finland. Gallant Finland fought alone during those months against the might of the Soviet Union. Finnish resistance was fierce in the winter of 1939, when in November the Soviet army invaded Finland and seized the province of Karelia in the southeast. The result was inevitable and Finland had to give up Karelia. War broke out again between the two countries in 1941 and lasted until the autumn of 1944. This time, with assistance from the Germans, the Finns managed to push the Soviet troops out of Karelia and back into Russian territory. But by 1944 the Soviet Union had captured Karelia for the second time. In the peace agreement which followed the war, Finland was forced to allow the Soviet Union to keep Karelia. They also had to give up the ice-free port of Petsamo (now called Pechenga) in the Arctic. This was unfortunate for Finland for, while the soil of Lapland is generally short of minerals, geologists had discovered seams of valuable nickel on the far north coast of the Arctic. The town of Petsamo had grown up almost overnight near the mines to cater for a mining population. Now that the Soviet Union claimed this area, Finland lost the nickel mines and the port and had no outlet into the Arctic Ocean.

The Soviet Union also demanded that the Finnish army should drive the German troops out of Finland. So the Finns began a campaign to expel the German troops from the northern and eastern parts of Finland. The Germans were furious about being turned out and they left behind them what has been called the most thorough devastation of any battlefield of the Second World

A street of wooden houses typical of many of the old towns destroyed in the Second World War.

War. The burned ruins of buildings, known as "Hitler's Monuments," covered Lapland which the Germans had destroyed in their retreat north through Norway. About 35,000 square miles (90,000 square kilometers) were laid waste. Almost every farmhouse, barn, telegraph pole, and haystack was destroyed in anger. That is why there are only a few buildings older than the 1940s in the whole of Finnish Lapland. Rovaniemi, the administrative center of the north (just south of the Arctic Circle), needed to be rebuilt completely and the Finnish architect Alvar Aalto was given that commission.

Finland was in a sorry state after the Second World War. During more than five years of war the Finns suffered 235,000 dead and wounded, a number equal to about 7 percent of the entire population, and a very high price to pay for defending its independence and freedom. The Soviet Union also suffered heavy losses in the Winter War when temperatures in eastern Finland, in an unusually cold winter, dropped to 40 degrees below zero. The Russian troops often froze to death or fell before Finland's ski troops dressed in white. The Finns could cope better with the climate, they carried fewer weapons and had small tents and wood stoves to keep themselves warm.

Peace with the Soviet Union was finally signed in 1947. The north was in ruins and Finland had to give to the Soviet Union 12 percent of its total territory. The Finns living in Karelia had the choice of remaining there, and so coming under Soviet rule, or of leaving their homes and going to live somewhere else in Finland. Almost all the Karelian Finns fled to Finland carrying a few possessions. Many burned their homes before they left to prevent the Russians from using them as shelter in the harsh conditions of winter. This created a great problem for the Finnish government who had to evacuate more than 400,000 Karelians and re-settle them in Finland. The Finns levied a tax on themselves to provide help and land for their dispossessed fellow countrymen. Finns from Karelia, the Porkkala area in the south, and from Petsamo in the north all had to be resettled. Between 1945 and 1950 Finland created 35,000 new farms to make homes for the Karelians, sometimes by slicing up large estates but mainly by starting new farms cut out of the forests. A new suburb was created in Lahti and a new town, Uusi Värtsilä, was built in the part of Karelia that was still Finnish.

Apart from having a great deal of rebuilding to do after the war, Finland had another problem. The thousands of Finnish children who had been sent to other Scandinavian countries (especially Sweden) for safety during the war now returned home not knowing how to speak their own language. Finland also had to pay a heavy war indemnity to the Soviet Union. It took it eight years to pay this off in the form of Finnish goods which were transported to the Soviet Union (an amount at first fixed at 300 million dollars, but later reduced). This huge payment was made without help from any other country.

Finland also had to give the Soviet Union a large part of its fleet of merchant ships. The naval base of Porkkala, west of Helsinki, had to be leased to the Soviet Union, supposedly for 50 years. During this time, because the railway line between the Finnish cities of Helsinki and Turku passed through Porkkala, the Finns in the trains were enclosed in coaches covered by iron shutters for this part of the route and it was made an offense to try to look outside. Only trains pulled by Russian engines were allowed to use the line which was often referred to by the Finns as their "longest tunnel." Porkkala was, however, handed back to Finland in 1956.

But for many years the Russian train that went daily from the Soviet Union through the south of Finland to Porkkala was a constant reminder of a lost war and the heavy price of defeat. But at least the war and the war reparations which had to be paid to the Soviet Union helped, by necessity, to improve Finnish industry and to expand the engineering and metal sectors of the economy; and at least Finland remained a united and an independent land. Since the war Finland has been on friendly terms with Russia. Finland's present policy is to cooperate and maintain friendly relations with all countries. It is a neutral country and does not take part in international military discussions.

Finland's armed forces are limited to internal duties by the Paris Peace Treaty of 1947. The numbers of troops allowed to serve in the armed forces are also limited. There are to be no more than 34,400 soldiers in the army, and this includes frontier troops and anti-aircraft artillery as well as the ski troops. The navy is limited to 4,500 sailors, and the air force to 3,000 and 60 aircraft. None of these aircraft are to be of a type designed especially for bombing.

Finland's neutrality means that it does not take part in military agreements such as the North Atlantic Treaty Organization (NATO). For Finland, being a neutral country means being independent politically of all other nations, but also seeking to cooperate with everyone.

4

Matters of State

Finland is a republic and the president of Finland is chosen to be head of state. He is a very powerful man. The president is chosen by an electoral college of 300 people (elected by the citizens of Finland) and he remains in office for six years. Urho Kekkonen's popularity resulted in several successive periods in the presidential office from 1956-1982. Finland's Scandinavian neighbors—Sweden, Norway, and Denmark—are monarchies with kings and queens as heads of state. But Finland has never been a monarchy and has had a president since the present constitution was adopted in 1919 when the country became a republic.

The president is powerful because he can stop a bill from becoming an act of parliament simply by refusing to sign it. He has the right of veto. But if there is then a general election and parliament approves the bill again without any amendments it will automatically become a law of the land. The president can issue rules and new laws, as long as these do not alter the existing law

The parliament building, Helsinki.

as agreed by parliament. He is commander-in-chief of the defense forces. He also appoints the Chancellor of Justice.

The parliament of Finland has exactly 200 members who are elected every four years as the result of a general election. The political party which wins the largest number of seats in parliament is, of course, the party which will rule the country. The most important politicians in this party form the cabinet of which the prime minister is the head. The members of the cabinet are appointed by the president.

There are about 11 political parties represented in the Finnish parliament, and several of them are parties with large numbers of members of parliament. The Finnish Social Democratic Party is rather similar to the Labor Party in Britain; the Finnish People's

32

Democratic League represents socialists and communists; the Center Party represents mainly the interests of people living in the countryside; the National Coalition Party is conservative; the Swedish People's Party represents the interests of the Swedish-speaking minority; the Finnish Rural Party often opposes the older parties; and there are other parties too. Finland's multi-party system means that the balance of political power does not often change very much from one election to another. In fact, since independence, no party has yet held an absolute majority in parliament. This means that most governments have been coalitions—agreements among the parties to govern jointly.

5

Helsinki—the Capital City

Helsinki (or Helsingfors to give the city its Swedish name) was founded by King Gustavus Vasa in 1550, when Finland was part of the kingdom of Sweden. But it was not until 1812 that it became Finland's capital city, taking the place of Turku. It was Czar Alexander I of Russia who decided that Helsinki should replace the ancient capital of Turku. He probably thought that Turku would always look toward Sweden and the west for its ideas, and that he could keep more control of Finland with a capital closer to Russia. In 1808, a fire destroyed parts of the old city and a large number of its wooden buildings. This gave the architect Carl Ludwig Engel the chance to make a fresh start in designing the city. He built a magnificent city center which is still admired today by all visitors to Helsinki.

During the summer the harbor is alive with boats—motor launches, pleasure boats packed with tourists, yachts, passenger ships from Copenhagen and Stockholm, cargo ships, and large passenger liners from all over the world. But in winter, much of the movement of boats ceases and only tugs are noticeable. Even these are mostly frozen into the ice in the harbor.

Helsinki is surrounded by sea and islands on three sides and in winter the city is only kept open as a harbor by the use of ice-breakers. These ships carry out winter patrols in the Gulf of Finland breaking a passage through the ice to allow cargo and passenger ships to enter and leave harbor. Large icebreakers keep a route permanently open between Stockholm, Turku, and Helsinki. Icebreakers are also needed to open routes for the first boats to make the journey each spring between Sweden and some of the smaller Finnish ports.

An aerial view of Helsinki, with the docks in the foreground.

The vegetable market in Helsinki.

This proximity of the sea to the very heart of the capital gives the city a spacious feeling. Parks and new buildings have been planned carefully to blend with those already there. Steamers from Sweden and Germany arrive in the very center of the city and moor in the South Harbor close to the market place. The ships that berth along the waterfront of Helsinki's harbor are one reason why Helsinki calls itself "the Daughter of the Baltic" and yet its trading and commercial ties are now with the whole world.

It is here that many of Finland's imports are unloaded. They range from machinery from the United States, to wool from Australia, fruit and vegetables from the Mediterranean lands, and cars from Europe and Japan. It is from here too that many Finnish exports leave the country, especially newsprint, timber, copper ingots, hand-finished fabrics, decorative glassware, and ships of all kinds, including huge freighters, cruisers, ice-breakers for use in the Baltic and in polar seas, and even oil-rigs, all of which are produced in the shipyards of Helsinki.

In the early morning fish, vegetables, fruit, berries, and flowers are sold in the market place with potatoes, peas, and fish sold directly from small boats. Near the market place is Engel's Senate Square where the huge Lutheran Cathedral, a beautiful white building with a green dome, sitting in the center of the square dominates the skyline. The university and government buildings are also in this square. Close by are the main post office, the railway station, the parliament building, the Finnish National Theater, and the main shopping streets and large department stores. The president's palace is by the harbor.

Although the center of Helsinki is compact, other parts of the

Helsinki Cathedral, sparkling white and with a green roof, dominates this square in the capital.

city are more far-flung and can only be reached by long walks or bus rides. The large children's hospital is known in Finland as the "Children's Castle" because it is 11-stories high. The Olympic Stadium was completed for the 1952 Olympic Games when Finland acted as host country. Outside the Olympic Stadium is a statue of the famed Finnish runner Paavo Nurmi designed by one of the greatest of Finnish sculptors Wäinö Aaltonen. In front of the National Theater there is a statue to the memory of Aleksis Kivi, the Finnish novelist who wrote *The Seven Brothers*, a story of Finnish boys growing up on a remote farm.

The city is a busy one all year-round, but it is a city of moods and contrasts. The winter snow and cold is replaced in the summer by sunbathing, swimming, and open air cafés. In June and August there are annual music festivals when Finland's greatest composer, Sibelius, is especially remembered.

Helsinki has the usual museums of any capital. The National Museum tells the story of the growth of the country in such areas as architecture, furniture, paintings, national costume, household utensils, and handicrafts. It shows clearly that Finnish culture has survived successive wars, invasions, fires, and plagues down the centuries.

There is also a museum in Helsinki devoted to the life and work of just one man, Carl Gustaf Mannerheim (1867-1951). He led the army in the civil war in 1918, in the troubled period following Finland's declaration of independence. In the Winter War of 1939 Mannerheim, then aged over 70, became commander-in-chief of the army. He could speak Finnish, Russian, English, French,

An attractive housing area in Tapiola. Notice the abundance of trees.

German, and Swedish, the last of which, as is the case with a minority of Finns, was his first tongue. Later, he became president of Finland and is commemorated today by the Mannerheimintie, an important street named after him in the capital where there is also a statue of him riding a horse.

Another museum particularly worth mentioning is the open-air museum on the island of Seurasaari. It consists of wooden buildings, some of them very old, which have been collected together from all over Finland. There are paths through the wood and clearings that contain old farms, houses, shops, a water mill, a granary, a windmill, barns, a sawmill, and even an old church. The museum has been growing since 1909. Squirrels dart to and fro across the paths and even climb visitors' legs to see if they are holding tidbits in their hands! Plays and folk dances in national costume are performed at this open-air museum during the summer; the old buildings give a magnificent setting to the performances.

The suburbs of Helsinki have been planned carefully as the city, now housing one in ten of Finland's population, has grown larger. In such well-planned suburbs as Tapiola, trees and rock have been left in place and houses have been built around them; this gives the feeling of a forest, even in the middle of a town. Housing, modern shops, schools, and churches of exciting new designs are all planned and built to a high standard. The Finns seem to have the happy knack of preserving the old, and also of making modern buildings seem different, exciting, and worthwhile. Planners and architects come from all parts of the world to see for themselves the garden suburbs of Finland.

6

Other Cities and Towns

Turku (or Abo in Swedish) was the capital of Finland until 1812 and is now the third largest city in the country. Turku has Finland's largest shipyards.

The old Handicraft Museum is on one of the seven hills of Turku in a part of the city which escaped the devastating fire of 1827. The wooden buildings in this open-air museum are exactly as they were in the old days, showing what a quaint place Turku must have been in the early 19th century. Women in the museum are dressed in traditional costumes and both men and women work at ancient handicrafts such as pottery making, tailoring, and spinning. There is a chemist, a carpenter, a clockmaker, a shoemaker, a printer, a bookbinder, a goldsmith, and a saddler. All these shops and craft rooms are equipped with the original tools of the trade. On one old building in the museum is a mirror placed at an angle outside a window. This was the mirror through which the "village gossip" kept an eye on everything that happened in the street, but without having to go outside to do so!

The wide Aura River runs through the center of Turku and its
banks are lined with small pleasure craft. There are tree-covered
walks and parks on either side of the embankment. The castle is
close to the harbor. It is an old one, begun in the 13th century,
and it includes some vast halls. Some of the castle rooms have
been converted into a museum. The cathedral, which was built in
the 13th and 14th centuries, is made of brick which seems strange
to anyone used to the stone-built cathedrals of Britain and western Europe.

Turku has modern blocks of apartments and new university
buildings alongside the remaining wooden buildings that have
survived intact from another age. In Finland the old and the new
blend well together.

The Abö Akademi is the only Finnish university in which the

42

teaching is purely in Swedish. About 6 percent of Finns speak Swedish as their first and native tongue. Books and newspapers are available in Swedish as well as Finnish editions.

Another distinctive building in Turku is the Resurrection Chapel in the cemetery. It is a world-famous chapel of very modern design that was designed by Erik Bryggman in 1941. Some of the walls are made entirely of glass through which the congregation in the chapel can see a glade of trees outside. The Finns are proud of their cemeteries which are always attractive places. All graves have flowers on them in the season when flowers are available.

Tampere (or Tammerfors in Swedish) is an inland city founded at the end of the 18th century. It is now the second largest city in the country and it is the main industrial city. Tampere is noted for

Tampere—Finland's main industrial city.

its modern buildings and for its clean and spacious appearance. The hundreds of factories are set amidst forests and lakes. In Tampere, as elsewhere in Finland, the industrial center is a clean place because electric power is used. Tampere is one of the leading textile cities in northern Europe, with large cotton spinning and weaving mills, and factories making knitted garments, linen goods, and synthetic textiles. Tampere also produces boots and shoes, as well as having flour mills and engineering works. The city is at the center of Finland's rubber industry.

Perhaps the most unusual sight in Tampere is the open-air theater in the Pyynikki National Park. The unique feature of this theater is the auditorium where the audience sit. It is a "bowl" holding 800 seats placed on a circular platform which moves round slowly between the scenes of a play. The audience sit still and the scenery changes as the bowl rotates to face another scene, ranging from forest to lake.

After the cities of Helsinki, Tampere, and Turku, Finland's towns seem much smaller. Many of these towns, such as Oulu, Pori, and Vaasa, are ports, as are Helsinki and Turku. A few towns such as Lahti and Kuopio are situated inland. Most other urban centers are quite small, such as Hämeenlinna in the south and Rovaniemi in the north. But, despite the size of these towns and cities, where about 60 percent of the population now works, the remaining 40 percent of Finns still live and work in the countryside, in villages and on farms.

7

The Countryside

Traveling northward by road or train in the late summer one passes hundreds of small farms with fields full of golden corn, and farmers working late into the evening to gather the precious grain. Most Finnish farms are small and keep animals as well as growing crops, although many farms have herds of less than 10 cows. About 8 percent of the working people in Finland earn their living from agriculture. The main crops are wheat, rye, barley, oats, potatoes, and sugar beet. But the harsh climate, the short growing season, and the danger of crop-killing arctic frosts in summer mean that most crops cannot do their best in such a northerly climate as that of Finland.

The climate does not affect animals as much as it does crops, and great progress has been made in dairy farming. Finland is able to grow almost enough grains for everyone, and it exports much dairy produce.

Timber is an extremely important world commodity. It provides not only wood but also paper, plastics, rayon, medicines, adhesives, dyes, and chemicals.

45

Forests cover more than seven-tenths of the land area of Finland—a greater percentage than in almost any other country. Yet, strangely, at times, Finland needs to import wood from Sweden and Russia to keep its mills working at full efficiency. The most important trees are pine, spruce, birch, and aspen. Pine, spruce, and birch are used in the timber and paper industries and aspen is turned into matches. Most of Finland's forests are owned by individual farmers and not by the government and so forestry becomes important and valuable to the farmer in winter when he is not able to till the ground. Since the late 1930s Finland has been one of the world's main exporters of wood and paper products.

As so many trees are cut down, the preservation of the forests for the needs of future generations must be constantly kept in mind. At present the annual rate of tree growth in Finland is greater than the annual rate of removal. Even so, conservation and reforestation, the caring for forests and the planting of new ones, is always an important consideration. Seeds are collected and young plants are grown in nurseries before being replanted as young trees in the forest.

A farmland scene, showing hay stacked on poles to dry.

Modern forest machinery stripping branches and foliage from the tree trunks.

Valkeakoski, near Tampere, is an important center of the timber industry. It is sited on a fall between two lakes and makes use of hydroelectric power. It produces paper, cellulose, and artificial fibers.

The trees that are destined to be cut down are marked at the end of the summer and are felled in the winter. After the branches have been stripped away, the trees are pulled by tractors or hauled over the snow on horse-drawn sledges, sometimes to a waiting lorry but more often to the banks of nearby rivers, streams, or lakes which are covered with ice. In the spring when the thaw comes the logs float downstream, through lakes and connecting canals, in huge bundles and rafts pulled by strong tugs. Logs may become jammed at bends in streams or even in a river. The lumberjacks use long poles to free the log jam. They jump from log to log (wearing spiked boots to stop them from slipping) to reach the trouble. This can be very dangerous work, especially when the river is a raging torrent of rushing water.

Timber is so important to the Finnish economy that trees are often referred to as the "green gold" of the country. When the logs are taken to the sawmills they are stored in huge ponds. The water protects the valuable timber from insects or fire. The logs are separated, depending on size and type of wood, and conveyor

A pulp mill.

Inside the paper mill.

belts are used to transport them into the sawmills. Then they are sprayed clean and the bark is stripped off. Often a machine is used to remove the bark. The logs are tumbled together in this machine until all the bark is torn off. Then the logs pass through saws which cut the timber into planks. The planks still contain much moisture and must be dried out or seasoned, either by stacking them in the open air or by using hot drying kilns. When the planks have been seasoned they are ready to be sent off for sale.

Timber that is destined to become paper is first made into wood pulp. It is either produced chemically or by machines in the paper

49

mills which chip the wood into pulp which is then washed, bleached and beaten. Chemicals are needed to make the paper shiny and to give it strength. The wet pulp is placed on a wire belt and the water drains away through the wire. The pulp is then squeezed through drums and rollers and the wet paper emerges. It is finally passed through a drying machine before being wound into very large rolls.

The products of the forest industries account for about one half of Finland's total exports to other countries. These products include timber, plywood, woodpulp, and newsprint. A daily newspaper that sells one million copies needs the newsprint that comes from over 2,000 trees every single day! But other kinds of paper—tissue paper, toilet paper, writing paper, birthday cards, wrapping paper, paper bags, drawing paper, and thick cardboard for book covers or cartons—are all made in much the same way. Thick cardboard is also used to make plasterboard which is used in the building of houses, both for ceilings and walls.

Life is still simple in many rural parts of Finland but modern living is catching up with even the remote hamlets. They now have electricity and mechanized farming as well as television and newspapers which spread ideas and reduce the feeling of isolation in small remote communities. Old customs last longer in the countryside than in the more sophisticated life of town dwellers. For instance, country dwellers visiting relatives living in a town indulge in the old Finnish habit of drinking seemingly endless cups of coffee while they sit and chat.

A lakeside village with its church spire peeping over the trees, and the surrounding farmland cultivated to the edge of the water.

Buses, cars, water buses, trains, and internal air services have all helped to make traveling easier for the Finns, whether they are visiting friends and relatives in towns or going on shopping sprees to the big city. But, like farmers the world over, country people are often tied to the farm by the sheer pressure of hard work and the necessity to milk cows regularly. For some country people in Finland, a visit to the nearby town is a welcome, but not necessarily a very frequent, pleasure.

8

Trout, Bears, and Cloudberries

Finland is a sparsely populated land of open unspoiled country-side, with vast numbers of trees, lakes, and islands. Bears, elk, and wolves roam in the remote areas of dense forests of pine, spruce, and birch trees. In winter, bears cross the Karelian frontier and enter Finland. A century ago they were much more common and even invaded the forests on the very outskirts of Helsinki in the winter months.

There are nearly 70 species of mammal native to Finland. Bears and wolves are now found only in the eastern part of the country and in the remote wastes of Lapland where there are also large herds of reindeer. Reindeer are domestic animals and each one is owned by somebody. At times, in the remote parts of Lapland, wolverines cause much damage to herds of reindeer. But, outside the areas of Lapland where reindeer are kept, it is an offense to kill wild animals. They are now protected in an attempt to save wildlife.

A herd of reindeer grazing. In spring and summer their antlers are protected with a velvet covering which rubs off when the antlers are fully grown.

Fur-bearing animals found in Finland include the muskrat, the red squirrel, the fox, and the pinemarten. Game birds such as wild duck, blackcock, ptarmigan, and grouse are common. Fishing is a popular pastime. Fishermen seek trout, grayling, pike, perch, and whitefish; and in Lapland there are many salmon in the rivers.

Wild flowers and berries are plentiful. The berries include red whortleberries, cranberries, bilberries, wild strawberries, and raspberries. In Lapland there are cloudberries which are an attractive orange color. Strong liqueur is made from the arctic cloudberry and varieties are eaten as a desert.

The lily-of-the-valley is Finland's national flower.

Education in Finland

During the 17th century the Church in Finland embarked on the task of teaching the nation to read. If parents could not teach their children to read, the local schoolmaster or some other literate person did so. In Aleksis Kivi's novel *The Seven Brothers*, set in the mid-19th century backwoods, the parson ordered the brothers to attend school to learn the alphabet. It is interesting that at the time the brothers were aged between 18 and 25! The first libraries in Finland were opened as early as 1850. The records for 1880 show that 97.6 percent of all inhabitants of Finland over 10 years of age were able to read and write. By the turn of the century, there were several thousand people's libraries to satisfy the widespread desire for books. It is well-known that Finland now has one of the highest rates of literacy, the ability to read and write, in the world.

During the second half of the 19th century, elementary schools were opened in all parts of Finland. Some of them were founded privately but others owed their beginnings to the churches. Often the original private schools were opened because state schools taught in Swedish. In fact, not until 1858 was the first Finnish-language school opened at Jyväskyla. Even today, the proportion of Swedish-language schools is higher than the six percent of Swedish-speaking Finns. Many local authority schools were the result of the first Elementary School Act of 1898 which made it the duty of the local authority to establish schools. After Finnish independence, the Compulsory School Attendance Act of 1921 required all children to attend school regularly for at least eight years.

The school year consists of two terms, with holidays at Christmas and in the summer (the latter being about 10 weeks long), a short skiing holiday in the spring, and one-day national holidays to celebrate Independence Day and May Day.

The school at Inari in Lapland where Lapp children sleep in dormitories during the school year.

An eight o'clock start in the morning is common in most Finnish schools. Children often arrive and also leave in the in darkness. It is especially difficult in country areas for children to get to and from school because of the harsh climate in winter and the long distances to be traveled to reach the school. The government pays for buses, taxis, and lodgings where children from distant farms live in the school year. Other children ski to school in winter, or go by boat or bicycle in the spring and autumn. Finnish schools used to work a six-day week, but they are now closed on Saturdays.

Some years ago, Finland changed its school system to the Basic School, or *peruskoulu*, a comprehensive school system. All pupils are now taught together in the same kind of community school for nine years (divided into schools for the various age ranges), between the ages of seven and 16. During the first six years, class teachers are responsible for most of the teaching of particular classes, but with specialist teaching in specific subjects gradually being introduced. In the last three years of the Basic School, specialist teachers of history, languages, mathematics and so on do most of the teaching.

Schools in Finland vary greatly in amenities and range from one-room log huts heated by wood-burning stoves to magnificent new buildings, many of them designed by famous Finnish architects.

Much time is spent on languages in Finnish schools. This is necessary because Finnish is of no use at all as an international language for trade and communication. English is the most popular

Finnish children dressed against the cold weather. The sign on the fence means "No Dogs."

foreign language to be studied in school, followed by German. Swedish is the second official language in the country and is taught as a matter of course in Finnish-language schools, just as Finnish is now taught in Swedish-language schools. Most schools start to teach a second language, English or Swedish, to nine-year-olds using oral methods and including such practical aids as language tapes, records, and workbooks.

At the age of 16, pupils can leave school or continue their studies for three further years in a senior secondary school, to prepare for university entrance. The comprehensive school for pupils aged between seven and 16 is free. This includes books and materials used in school, school meals, health care, and also transportation to school when necessary. In the senior secondary schools pupils usually have to pay for tuition, although some local

57

authorities help parents by providing some of the money, and there are always free places for pupils whose parents cannot afford tuition fees. About half the 16-year-olds each year enter senior secondary schools, a very high proportion. The subjects that are taught in the senior secondary schools, as in the upper forms of the Basic School, are divided into compulsory subjects (which everyone studies) and optional subjects from which each pupil chooses several subjects that he would like to study.

Other pupils may go at 16 to a vocational school where they learn some of the practical skills needed in the job they intend to follow as their career. The Mustiala agricultural school opened as far back as 1840. There are vocational schools and colleges in agriculture and forestry and, in addition, students choose from a great many types of courses—trades, office work, sales promotion, transportation, safety, public health, nursing, and many more.

School pupils who are accepted for entrance to a university become students at about 19 years of age. Students wear the student cap that is so proudly displayed throughout Scandinavia. Exam pressures are high and Finnish students are very conscious of their marks, remembering clearly many years later what marks they achieved in important examinations. There is a high standard at Finnish universities. Four or five years is the minimum period of study for a degree. Many degrees take much longer. There are equal numbers of male and female students. Many Finnish women follow specialist careers; most Finnish dentists are women and there are many women engineers and architects. Women have always had an important place in Finnish society, and Finland was the first country in Europe to give women the right to vote in elections. They cast their votes for the first time in 1907.

Finnish students, with such long studies before them, take on jobs during the holidays and sometimes during term as well. Many students who cannot get financial help from their parents have to work their way through university. One way of financing study is by obtaining a state study loan with low interest rates which has to be paid back within a specified number of years.

There are nearly 20 universities or colleges of higher education in Finland. The University of Helsinki is the largest in Scandinavia and one of the largest in all Europe. There are places at universities and colleges for about 12,000 new students each year, and this is about half the total number of pupils who leave senior secondary schools each year. The largest universities are at Helsinki, Turku, Tampere, Jyväskylä, and Oulu. Institutions of higher learning include the Helsinki University of Technology and the Helsinki School of Economics and Business Administration.

Education for adults takes many forms in Finland. There are

Part of Turku University.

people's colleges offering adult education, workers' evening schools, and people's institutes. There are several thousand study circles, as well as correspondence schools with large numbers taking postal courses.

10

Games and Sports

Most Finns are very keen on games and sports of all kinds, especially athletics and winter sports. In summer there are track and field events and soccer. In winter there is skiing and ski jumping. A great many Finns not only enjoy watching sports as spectators but also actively take part, either for enjoyment or just to keep themselves fit. Many play ball games and participate in track and field events, and skiing. Others enjoy gymnastics, hiking, swim-

Skiing is a popular pastime with young and old.

ming, wrestling, boxing, skating, or cycling. Local ski tracks, swimming pools, gyms, and indoor sports stadiums are always popular and often in use.

Skiing is the national pastime of the Finns. Today, large numbers of Finns in all parts of the country ski for enjoyment. In the past, before communications improved, it was necessary to be able to ski to get from one village to another in winter. Cross-country skiing is very popular and more enjoyable than merely skiing down the same slope and being carried up on a ski lift to ski down yet again. There are very few ski lifts in Finland.

Ski jumping is also very popular. Finland has won many Olympic awards for this sport. Many villages and almost every town has its ski jump and there are more than 200 of them in Finland. A large one at Rukatunturi, near Kuusamo close to the Arctic Circle, is used for jumps up to 115 meters. The Rovaniemi International Winter Games are held at the 90-meter ski jump. Other international events are the winter games at both Lahti and Kuopio.

Finland has two sports especially its own. One is *pesäpallo*, a form of baseball which may be described as the national game of the Finns. It is played enthusiastically by children from an early age. Another is *jääkiekko*, rather like hockey but played on ice.

Sailing is a common sport in the summer and great use is made of both the lakes and the sea. There is an annual Helsinki Regatta in the Gulf of Finland. The sloops which compete have plastic windows in their sails so that the yachtsmen can see through them to help prevent collision with the other boats in the race. Helsinki,

Turku, and Hanko are the venues for major yachting attractions, although there are many smaller events held around the coast and also inland.

The lakes and rivers of Finland are ideal for canoeing, and canoe rallies are also held in Lapland where there are spectacular rapids.

Finns compete keenly in international competitions. One of the greatest sporting names Finland ever produced was Paavo Nurmi who, from the 1920s, collected nine gold medals during his track career as a long-distance runner.

A more recent name with international repute is that of Lasse Viren who earned two gold medals in track events (5,000 meters and 10,000 meters) in the 1972 Munich games and repeated the performance in the 1976 Montreal games. The first World Athletics Championships were held in Helsinki in 1983.

Finns also often hit the international headlines when competing in racing-car and rally events.

11

The Sauna

Most Finnish homes contain a bathroom and a sauna, a small room lined with pine wood in which you become very hot and also very clean. Inside the sauna there is a large stove which has a chimney to take the smoke outside. Long before the bather enters the sauna, a fire in the stove makes the room very hot. The fire may burn wood or the stove may be heated by electricity.

You leave your clothes in the changing room and then go into the sauna, closing the door behind you. Inside there are wooden benches, rising in several tiers. The stove is hot and soon you begin to feel extremely hot yourself. Close to where you are sitting you have a bowl or a bucket with cold water in it. You use a scoop to throw some water on the hot stones on top of the stove. There is a hissing noise and you sit back and wait for the dry heat to hit you! The whole room becomes filled with a cloud of steam as the water evaporates. It is best not to breathe in just at that point. The steam is so amazingly hot that for a few seconds it is difficult to breathe properly. You now begin to sweat a great deal. Then the heat becomes less severe again.

When you have been in the sauna for perhaps up to half an hour you usually go into the changing room to have a cool shower. You feel so hot that even a very cold shower seems only mildly warm, and the cold water gives you a tingling feeling.

After resting for a short while you can go back into the sauna and start to get hot all over again! Usually the Finns have a sauna once a week, often on Saturday which is the traditional sauna day, but many people go more often. Children love to have a sauna and they start at a very early age.

The hottest part of the sauna is near the ceiling. An experienced sauna bather can sit or lie on the wooden bench close to the ceiling and stay there for a long time, even though it is so very hot. But beginners are advised to sit on a bench close to the floor as that will be quite hot enough for them.

The temperature in the sauna is often around 212 degrees Fahrenheit (100 degrees Celsius). The bather does not melt away under such heat because the human body can stand a great deal of this dry heat. Saunas are built of wood or lined inside with wood. The wood does not become too hot to touch. Only the stove is dangerous to touch and small children are warned not to go near it.

Often you are not alone in the sauna. Sometimes a father goes in with his sons, and a mother and her daughters go in later. Sometimes the parents go in together and the children go in together later. Sometimes the whole family may go in together on sauna night. Most saunas are large enough for everyone to squeeze in. The sauna is more than just a way of becoming clean. It is the national habit of most Finns. The sauna is almost as nec-

A sauna—the stove is on the left. Note the birch twigs in the wooden tub.

essary to the Finns as food and drink. In the past, when a farmer and his family settled on new land they often built their sauna before they built their farmhouse. They lived in the sauna while their farmhouse was being built.

Not every family in Finland owns a sauna but many families do have one. There are nearly a million saunas shared among a population of five million. Finnish families who do not have their own often share a sauna belonging to relatives or friends. If a family lives in a block of apartments, they use the communal saunas in the basement; there will be a certain hour in the week when a sauna is reserved for the exclusive use of each family. There are some public saunas. In addition, many factories and offices have a sauna in the building which is used by the people who work there.

Inside a sauna it is possible to feel so relaxed and drowsy that you may fall asleep. An experienced sauna user likes to be beaten with birch twigs. The twigs have the leaves still on them. Almost every sauna has a bundle of birch twigs hanging in a corner. These twigs are sold in markets in Finland for use in saunas. One bather may beat another quite hard on his back. The beating helps the circulation of blood through the body and makes you feel very refreshed. It does not hurt the skin because the twigs are soft and supple and they bend easily. The birch twigs are soaked in a bucket of water to soften the leaves before they are used. Bathers take turns at beating each other on the back. If you are in a sauna by yourself you can beat your own arms, legs, and body with the twigs. Although birch twigs are common in saunas, some Finns do not bother to use them at all.

After you have been sitting for about half an hour in a sauna the pores of your skin are cleaned really well. When you leave the sauna for the last time you should lie down and rest in a room of ordinary temperature to let your body temperature return to normal. For the beginner, having a sauna is an exciting but also rather a tiring experience; some look red all over and are ready for a cold drink and then bed. Others, after a rest period, feel fit enough to conquer the world!

One of the highest honors a Finn can offer his guest is to invite him to share his sauna. Many Finns invite their friends to have a sauna in the same way that people in other countries invite friends in for a meal or for a drink. Friends sit in the sauna and relax or talk and joke quietly. This may go on for several hours, with showers and rests every so often. An average-sized sauna may hold about a dozen people, although others are much smaller.

Not all saunas are erected inside buildings. Many thousands of them are simply small log or wooden huts built beside the sea or on the banks of a river. Others are built on the shores of Finland's lakes or islands. Usually, these saunas have two small rooms. The outer room is where you leave your clothes. The inner room is the sauna containing the stove and the long wooden benches to sit on.

Many Finnish families have two homes—their ordinary home and also a small cottage beside a lake or the sea which they use in the summer for holidays. If they can afford it, they also have two saunas, one in each home. When you are ready to cool off from

Many Finns take their vacations in beautiful surroundings like this, with forests and a lake nearby.

a sauna beside a summer cottage you simply run outside and plunge into the water whether it be a lake, a river, or the sea. Many of these saunas are built in quiet, secluded spots, but even so most bathers usually peep round the door of the sauna to see if anyone is passing in a boat before they run out naked to jump into the water! In winter, some Finns still use saunas built in lonely places. They roll about naked in the snow to cool themselves and do not feel in the least cold when doing so.

One amusing story is told about an official of a relief organization from another country who visited northern Finland in the gloom of winter just after the Second World War. His job was to find out which things the people needed and to try to get supplies to them. He had obviously never heard of a sauna in his life for he was surprised and horrified to see a naked family suddenly run out of a very small wooden building and begin to jump about and roll around in the snow. He sent an urgent message back to his headquarters: "Send these people supplies of clothing quickly. They live in tiny wooden huts. They have no clothes at all and they have to run about in the snow to try to keep themselves warm." He would have been surprised had he known that they were rolling in the snow to make themselves cool!

The sauna is very much a way of life in Finland. It is part of the weekly ritual of the poor and the rich, of the peasant and the president. The president of Finland has large saunas in his official residences and in these he sometimes entertains his guests. On one occasion President Urho Kekkonen of Finland discussed important government plans in a sauna with the premier of Russia.

A sauna built on the edge of the Baltic.

In earlier times, and even in some country saunas today, the sauna did not have a chimney and was heated by the smoke circulating from the stove. The bathers had to wait until the smoke disappeared through the door before they could enter. Apart from taking longer to get ready for use, this older type of sauna made the bather's eyes sting because some smoke was sure to be left inside. There are some saunas of this kind still in use in the countryside and some people prefer them as they like the smell of the pine-wood smoke.

In the north of Finland the Lapps who have settled there permanently are as sauna-minded as the Finns. Other countries have also copied the idea of building saunas. Many of the German troops who were stationed in Finland during the Second World War built saunas in Germany when they returned home. There are some public saunas in Britain and the United States, attached to swimming pools, and a few firms specialize in building saunas in houses for the use of individual families.

12

Lapland

It is a long journey by train from the south of Finland to Lapland in the north. From Turku, for instance, there is a rail journey of over two hours to Toijala where you change trains. The second part of the journey is in a luxury train from Helsinki with reclining seats which are needed for the 12-hour journey to Kemi. (Or, if you are traveling by night, you can book a sleeper.) At Kemi another train takes the traveler to Rovaniemi. This journey, on a single-track line, takes a further two and a half hours. From Rovaniemi a fast bus goes north to Inari in the center of Finnish Lapland, but the bus journey takes seven and a half hours!

The Lapps are a people who live in the north of Norway, Sweden, and Finland and in the Kola peninsula in Russia. Lapland is not the name of a country; it is the name of the northern part of Scandinavia. Many Lapps still have their traditional tunics, leggings, and colorful hats but these are not usually worn all the time. Many wear shoes and waterproof boots bought from the nearest store instead of making their own moccasin shoes

Lapps wearing traditional dress outside the church—but notice the girl's modern plastic bag.

from reindeer skin, and some prefer to buy clothes ready-made instead of making their own. Usually traditional dress is now worn only at festivals or on important occasions such as weddings.

The Lapps in Finnish Lapland are vastly outnumbered by the Finns. Many Lapps have settled down and live in small villages. The men may find work in mining, seasonal agriculture, felling trees, or fishing. Some Lapps, however, still continue to live as their ancestors did, herding reindeer. They live very simply and follow the reindeer when they search for food. In summer, the reindeer feed on the fells, partly because the breeze keeps away the irritating mosquitoes. In winter, the reindeer move to the forests in the south of Lapland where the snow is not so dense as on the frozen tundra. Reindeer eat lichen and moss and in the winter they must use their antlers and hooves to paw away the snow in order to get food to keep themselves alive. In winter, they drink by eating mouthfuls of snow.

The Lapp who keeps reindeer must be prepared to travel a great deal. At one time the whole family followed the reindeer but today usually only the men and youths go on winter migration. The women and children stay in a small log hut, usually with only one room. While watching the herds in winter, the men live in a small lightweight tent, smaller than the tent used in summer when all the

A winter round-up of reindeer in one of many Lapp corrals.

A Lapp tent used by a herding family in summer.

family will be living in it. The reindeer are guarded, even at night, in case prowling wolves come too near. Bears are now rare but the Lapp on guard may have a rifle just in case a bear approaches. He has a trained dog to help him control and protect the herd.

Lapp children grow up very quickly as they have to be trusted to do all kinds of responsible jobs. They have few toys and their main game is playing at lassoing reindeer. This, for the boys at least, will be very important in their lives as adults. Reindeer have to be lassoed and counted at round-ups during the winter and in the spring when some are slaughtered for meat and the young ones have their ears notched with their owner's identification marks.

74

At one time the reindeer gave the Lapp a home, tools, food, milk, and clothes. The tent was made from reindeer hides, as was clothing. Tools were made from antlers. Reindeer meat was the most common food available in Lapland. In the Lappish language there are more than 20 different words for reindeer—a large reindeer, a bull, a calf and so on, showing how important reindeer are to the Lapps. But now that other foods and clothes can be easily bought in shops the reindeer have lost some of their former importance. Even so, they are still vital to the herding Lapp who needs at the very least 200 reindeer to support his family. In winter, the Lapps load their possessions onto sleds that are pulled by reindeer. Reindeer lose their antlers in winter and grow new ones in the spring, so they tend to look rather bare and forlorn once these wonderful head adornments have been shed. They are the only species of deer in which the female as well as the male has antlers. But the antlers of the male are always larger and more magnificent. In addition to antlers, the male has a brow-tine which helps it to scrape away snow in winter in the search for moss and lichens. A buck's antlers grow in the spring and are shed by older bucks in November; only the young animals and the females retain their antlers into the new year.

The Lapps are now able to buy snow scooters (rather like motorized sleds) which can pull great loads. Traditionalists prefer to use reindeer to carry and pull their belongings. Others have bought these scooters for winter use, although they are expensive, since they do the work of 20 reindeer. Tractor buses travel around Lapland in winter. They can move over ground which ordinary vehicles would find impossible in the winter snow. These special snow buses carry 20 passengers and look rather like a tank on runners. The Lapps are also very good at skiing. Most children have skis that are both fun to use and a necessary way of traveling in winter.

In the summer when the sun shines Lapland can be a very warm place, very different from the dark and cold region of the winter months. Winters are bitterly cold in Lapland and everyone wraps up well and wears thick gloves. If they did not wear gloves their fingers would be frostbitten. The felt trousers of summer are changed in the winter for thick leggings made from reindeer skin. And everyone puts on even more clothes at night. They never undress for bed as we do.

The Lapps are a small people and are very sturdy. They have short noses and dark hair. Generally, they are happy people living a peaceful, simple life. For them, food, warmth, and the welfare of their reindeer are the only really important things in life.

At Inari there is an open-air museum featuring different kinds of Lapp homes. The museum shows methods of building with logs, a nomad village, reindeer stables, a fishing village, and an exhibition of handicrafts, clothes, and wooden household utensils. (Most things in a Lapp home are made from wood.)

Some Lapp children who live too far from the nearest school to travel there daily, must board at the school during the school year. Others live close enough to school to live at home and walk or ski to school, or ride on a *pulkka*—a small sled with one runner, rather like a canoe, that is pulled by a reindeer.

The Lapps have a long history. In fact one old Lapp ski has been discovered and dated as being 3,500 years old. At home the Lapps in Finland speak one of the three dialects of the ancient Lappish language which is a mixture of Hungarian and Finnish, but at school the children learn to speak Finnish since they are citizens of Finland. They learn to read, write, and do number work, but they are also taught the things they need to know when they grow up—how to care for sick reindeer, about ear identification markings, and how to sew and weave.

Everyone in Lapland looks forward to Easter when Lapps living in remote regions visit the nearest village, wearing their best clothes. They come to do shopping and to sell a few reindeer. They also meet their nearest "neighbors" whom they may not have seen since the previous spring! They worship at the small wooden church in the village, and weddings and baptisms take place. A Lapp wedding is a very colorful sight as all the guests wear beautiful traditional clothes of the brightest colors. Spring comes quickly in Lapland and flowers bloom on ground which was covered with ice a few days before. Everyone is relaxed and glad that the cold and difficult winter is over. Sleds are pulled by reindeer wearing decorations on their harness. When the Lapps get together at these times they talk a lot and sing together to help make up for the lonely life they usually have. They hold competitions including the lassoing of reindeer to see who is the most skilled. Even when they are enjoying themselves the Lapps find fun and amusement with their reindeer.

The nomadic Lapp prefers the quiet way of life in the lonely

A reconstruction of an old Lapp timber building. It would originally have been covered with turf to keep out drafts.

hills. He is a little like his reindeer. They are both rather shy and each feels happier when among the lonely hills. Some small hamlets and villages in Lapland are in extremely lonely places and in winter they are cut off as soon as the snow falls. When there is an emergency a helicopter may fly in to carry an injured person to the hospital. Telephones, planes that land on lakes, helicopters, snow scooters and snow buses mean that Lapland is no longer quite so cut off and snowbound during the winter months.

The governments of the Scandinavian countries, including Finland, have spent much money on building fine new roads which run straight through Lapland, and along which flow an increasing number of cars and other vehicles, including post buses. These are ordinary buses carrying passengers but they also drop off letters, packages, and newspapers for people who live along their routes. Special road signs in Lapland carry a picture of a reindeer and warn of reindeer crossing the road. At night this is a particular hazard, for a reindeer crossing a road will usually run straight toward the bright headlights of a car.

Lapland has been described as one of the most beautiful and peaceful borders in the world. The long light nights with the Midnight Sun in summer, the blaze of autumn color on the fells, and the red, orange, green and yellow of the Northern Lights in the sky during the months of winter darkness are some of the special features of Lapland. Increasing numbers of skiers and anglers now visit Lapland for sporting vacations.

Every year more and more Lapps find jobs in the villages and small towns in the north and so fewer of them are herding reindeer. The Lapps are one of the oldest European tribes. They have looked after reindeer for thousands of years. But the life of the nomad is a very tough one and some Lapps feel that they can earn their living easier by doing other work. An old Lapp saying is that the Lapp is born to suffer as the bird is born to fly. It is true that the nomad needs to be very strong and extremely fit. Visitors now go on vacation to Lapland from other countries and from the south of Finland. They take their cameras with them and they stare curiously at the Lapps in their unusual bright costumes. The Lapps, of course, do not like being stared at, although those in popular tourist centers have grown used to it by now. Those who live in expanding villages are having to change their attitudes and their lifestyles. Some now work in the tourist industry at vacation centers where there are illuminated slalom slopes (ski courses with artificial obstacles), reindeer-driving schools, reindeer safaris in winter (traveling with the Lapps for two or three days, crossing the fells on sleds and staying at huts for meals and at night), and even washing for gold on the Lemmenjoki River and the Ivalojoki River. But probably there will always be some Lapps who herd reindeer, for this is what the Lapp is happiest doing and this, from the beginning of history, is what he has always done.

13

Living in Finland

A meal in Finland often consists of fish soup and a "cold table," as in all Scandinavia, including such foods as fish, bread (especially rye bread), boiled potato, cold meats including smoked reindeer, or rolls with cheese, cucumber, and tomatoes. Yogurt is often eaten too. Hot meals include Karelian pasty (made of rye flour) filled with potato or rice, fish and pork pie, hot-pot of veal, pork, and mutton, boiled fish in sauce, and braised reindeer stew. A number of dishes from Russia are often eaten such as beef stroganoff, beetroot soup with sour cream, and pancakes with sour cream and onions. Many Finns drink a great deal of milk with their meals, and mild beer is also popular.

It is difficult to make a general statement about a typical Finnish home for people live in homes of all kinds, ranging from an apartment in a high rise to old wooden buildings and modern detached homes. In some houses the main rooms are not on the ground floor but have a garage, storeroom, sauna, and central-heating system below them in the semi-basement. The main living rooms in this kind of house are higher off the ground than we are used to, and the bedrooms are in the roof space. Winters in Finland are so very cold that most houses have central-heating systems. Finland still uses a great deal of wood for heating homes. Windows are double- or triple-glazed to keep out the cold. (Even the windows in buses and trains are double-glazed.)

Yards may be large or small with lawns, rose bushes, apple trees, strawberries, gooseberry, blackcurrant bushes, and perhaps a greenhouse growing tomatoes and cucumbers. Many gardens are very attractive in the summer and have colorful displays of flowers. As the summer lasts for only a few months before the cruel snow of winter falls, the Finns try to grow as many flowers as they can during the summer. In winter, flowers must be grown indoors and all families have pot plants. Flowers are difficult to

A house built on three levels—sauna, garage, and workroom in the basement; bedrooms in the roof; and living rooms between.

A Finnish brother and sister in a rowing boat near their summer cottage.

buy during the long winters. In fact, a Finnish visitor may take his host a single flower as a present, attractively displayed inside a polythene wrapper.

In summer, the Finns wear the same clothes as are worn in western Europe. But in winter very warm clothing is needed by everyone and it includes long woolen underwear, gloves, and protective caps and hats. In extremely cold weather nylon stockings might freeze to the wearer's legs!

The Finns have different clothing for different seasons. This is because of the extremes in climate and temperature at different times of the year. In summer the temperature often reaches 82 degrees Fahrenheit (28 degrees Celsius) with long sunny days, thanks to the mild influence of the Gulf Stream from the west and to the dry climate from continental Europe in the east. But winter starts in November and the first snow often falls in December. The winter months until January are dark and depressing. In January and February the temperature in the north is often as low as *minus* 13 degrees Fahrenheit (*minus* 25 degrees Celsius) and at times even *minus* 40 degrees Fahrenheit (*minus* 40 degrees Celsius). In the south it falls to five degrees Fahrenheit (*minus* 15 degrees Celsius). But by March there are up to 16 hours of daylight in the south. The thaw begins in April and by the start of May the snow has all disappeared in the south although it goes about a month later in Lapland.

Almost everyone in Finland owns or rents a second home, usually by the side of a lake, a river, or the sea. The family spends the summer at this second home which may be just a one-room chalet or a large cottage. Finnish children have about 10 weeks for their summer vacations. Many go with their mother to a summer cottage when the school term ends in early June, often staying until the middle of August. Father joins the family at weekends, or even each evening if he does not have too far to travel. Then he comes and stays full-time when his own vacation comes round. Some of the inhabitants of Turku, for instance, have summer cottages on the island of Merimasku, and to reach one of these island cottages from Turku may entail a bus ride of over an hour, crossing the bridge to the island, and then a long walk.

Some cottages are about the size of an ordinary house with several rooms, including bedrooms upstairs. But summer cottages, because of their inaccessibility to mains services, often lack the refinements of a town house. Water is often drawn from a well. There may not be electricity so candles or oil lamps are necessary in the evenings, and cooking on a primus or woodburning stove must be relatively simple. Such cottages often do not have running water either so the Finns wash themselves in a lake, or river, or in the Baltic, and washing-up after a meal is done in the same way. The lavatory is often in a wooden shed in the forest.

Most cottages are far enough away from other cottages to be secluded and peaceful. Milk is obtained from the nearest farmer and supplies of food from the nearest village shop. The Finns enjoy this change of surroundings in the summer months when they can swim, walk, boat, fish, and pick berries. Many cottages have a rowing boat and perhaps even a dock. Some families place a net in the water and row out each morning and evening to remove the fish that have been caught. Then they make fish soup.

Mosquitoes are troublesome in Finland from June until August. Some cottage windows are covered inside with a fine wire mesh to prevent the mosquitoes from entering. This insect is the greatest nuisance in Lapland. Leaving an outside door open for only a few seconds is long enough to invite mosquitoes to enter. When they fly they make a high-pitched buzz. Once you have been bitten by them the irritation and the temptation to scratch is very great, resulting in red patches on the skin for a few days.

Mariehamn is the capital of the Aland Islands owned by Finland. These islands lie between Stockholm and Turku. The inhabitants speak Swedish. The group of about 6,500 Aland Islands has a character of its own. Fishing there is for cod and

A summer cottage. The colors of the Finnish flag represent blue sky and white snow.

flounder that is the chief occupation on the islands. Deep in winter, the islanders cut holes in the ice of the frozen sea and fish through the holes. Many of the islands are wind-swept rocks without trees and with little land capable of cultivation Other areas have good soil where farming is important. Some islands near Turku are now used by Finns from the mainland who have summer cottages there.

The standard of communications between one part of Finland and another is high, considering the long distances involved, the problems caused by the many lakes, the long periods during which some roads are impassable in winter, and the problems caused by frozen harbors. Buses travel to all parts of the country carrying people and mail. Traffic drives on the right in Finland, as is common in Europe and the United States. Roads have been improved in recent years but surfaces sometimes become damaged after the snow and ice of winter. There are well-surfaced roads between the main cities and towns. But many others are dirt roads, dusty in dry weather, muddy in spring, and covered with hard-packed snow in winter.

Finland has a modern rail system with about 5,500 miles (9,000 kilometers) of track which often has to make a detour around a lake. The railway network began in 1862 with the opening of the first line in Finland between Helsinki and Hämeenlinna. British rolling stock was used from Birmingham and Birkenhead and two British engine-drivers stayed for a few years to instruct the Finns. The railway was built to make use of the timber of the region around the westerly group of lakes close to Hämeenlinna. The old engines were wood-fired, then came coal-fired steam engines; but today diesel trains are common everywhere and part of the rail network is electric. Trains transport freight, and carry passengers between the main cities and also between cities and their suburbs.

Air travel is important in Finland. The most important aviation company is called Finnair. This is mainly owned by the Finnish State. Apart from the normal passenger services, charter flights can be arranged for groups of passengers who hire an aircraft. It

Repairing the surface of the road which often cracks in severe winters.

is possible to fly from Helsinki to Rovaniemi in only one and a half hours. The most northerly airport in the country is at Ivalo in Lapland. There are also many local flights—between Turku and Helsinki, for instance, even though this distance can be covered by train in about three hours. Even a short flight shows very clearly from the air the extent of Finland's innumerable lakes and vast tracts of forest. Finnair has services to the other Scandinavian countries, to Russia, to New York, and to many parts of Europe and Asia.

Lake steamers serve as water buses, carrying passengers around the interconnecting lakes. Some are modern hydrofoil services, while others are old wood-burning steamers built at the beginning of the century. Some steamers have restaurants on board and are popular with tourists because they visit scenic parts of Finland. The steamers on the huge Lake Saimaa follow routes which serve local people and also tourists who may cruise for a week on the lake. They sleep on board at night and visit such towns as

One of the many water buses which carry passengers from place to place around the inter-connecting system of lakes.

Lappeenranta, Savonlinna, Joensuu, and Kuopio as well as stopping at isolated villages.

There are also coastal steamers, especially between Turku and the Aland Islands. There is a regular service between Helsinki and Porvoo, and a motorboat from Kotka to the vacation island of Kaunissaari. Silja Lines have modern car ferries which ply a direct sea route from Travemünde in Germany to Helsinki in 22 hours. There are also services from Helsinki to Tallinn in Russia. Many car-ferry services run daily between Sweden and Finland. The Viking Line, for instance, has services from Stockholm to Turku, Naantali (where the president has a summer palace), and Helsinki. They may also call at Mariehamn on the way.

14

Finland in the Modern World

Finland is now one of the richest countries in the world with a high standard of living. Since independence Finland has changed from an agricultural nation into an industrial nation. Early in the 20th century Finnish industry was mainly concerned with wood-processing, but the Second World War saw the growth of the metal and engineering industries.

The economy of Finland depends on foreign trade. Some of Finland's most important trading partners are Russia, Sweden, Britain, and Germany as well as Japan and the United States. Naturally enough, Finland has a close trading relationship with the other Scandinavian countries of Denmark, Norway, Sweden, and Iceland. In 1962 these five countries made an agreement, called the Helsinki Treaty, aimed at strengthening and expanding cooperation in Scandinavia. In 1975 the Scandinavian countries established a Nordic Investment Bank with its headquarters in

A Finnish passenger ship arriving at Turku from Stockholm after a night in the Baltic in summer.

Helsinki. This bank encourages Scandinavian investment and exports by making loans.

Relations between Finland and Russia improved. A treaty of friendship, cooperation, and mutual assistance signed in 1948 was renewed in 1970 and 1983. There were joint projects in technical and industrial cooperation, such as the construction of forestry and metal industry centers. Finland supplied copper smelting equipment to Russia for a plant in Siberia. Similarly, Russia supplied Finland with atomic power plants. Finland's first nuclear power plant was opened at Loviisa in 1977 to help to satisfy part of the country's increasing demand for power. The Pirttikoski power station on the Kemi River in the north is built entirely underground, the machine chamber having been blasted out of solid rock. One of Finland's largest steel mills, at Raahe on the Gulf of Bothnia, is a cooperative undertaking with Russia. Finland exports to Russia forest industry machinery, paper, and ships and imports industrial raw materials, fuels such as natural gas, automobiles, and machinery.

Russia leased its part of the Saimaa Canal to Finland from 1968. This waterway is important because it connects the inland waters of Finland with the Gulf of Finland and the Baltic. This makes it much easier to ship timber and paper abroad as large ships are able to use the canal. The leasing was seen as an act of goodwill in Finnish-Russian relations. Finnish-Russian tourism also increasing and Russian vacationers became common in Finland.

In 1990 the cooperation between Russia and Finland broke down for a while. Finland's economy went into a recession from which it only began to recover in 1992. Finland joined the European Union in 1995 to try to protect its economy and reduce unemployment by increasing trade with other European Union countries.

The metal industry now employs more workers than any other industry in the country. One Finn in every three working in industry is employed by the metal industry. Finland is one of the most important producers of copper and nickel in Europe. Other ores include zinc, iron, chromium, cobalt, and vanadium. Finland produces ships, cars, lorries, locomotives, railway rolling stock, cranes, earth-moving machines, agricultural machines, and wood-working industry machinery.

The Finnish chemical industry has grown greatly in recent years, producing raw chemicals, fertilizers, plastics, medicines, paints, dyes, soap, fuels, and lubricants. Finnish scientists have developed xylitol which is a sweetening substance obtained from birch. Its great advantage is that it does not cause tooth decay.

In recent times Finland, in common with other industrial nations, has had many economic problems. These were part of the economic crisis affecting countries all over the world, causing unemployment, inflation, and the slowing down of economic growth. The Finnish government actively encourages investment and the competitive efficiency of businesses.

The Suomen Pankki (Bank of Finland) prints money and supervises the monetary system under the control of parliament. The unit of currency in Finland is the *markka* and 100 *penniä* are worth one *markka*. In other countries Finnish currency is referred to as the Finnmark (abbreviated to Fmk) to distinguish it from the marks of other nations. There are notes for the values of five, ten, one hundred, and five hundred Finnmarks, and coins for the denominations five, ten, twenty, and fifty penniä and also for one Finnmark.

Tourism is an increasingly important service industry in Finland and, as it brings foreign money into the country, tourism is counted as an "export" industry. This includes employment in hotels and restaurants, and transporting visitors around the country. At present, about three million foreign tourists visit Finland each year; many of them are Scandinavians, and most of these are Swedes. The tourist slogan for Finland is "Four Seasons—Four Reasons," stressing that Finland can be visited at any time of year, and that each season is different: sun in summer, autumn colors as trees change, skiing, theaters, and other cultural events in winter, and a short colorful spring in Lapland with skiing there in sunshine until May.

The foreign visitor will find many Finns able to speak English which is just as well because Finnish is such a difficult language to learn. It is related to Hungarian and Estonian. In pronouncing a Finnish word the stress usually falls on the first syllable and it is a phonetic language—the words sound the way they are spelled. But short or long vowels can cause problems. For example, *vaja* means "shed," *vaaja* means "wedge," and *vajaa* means "scant." A few words in Finnish will give you a taste of how difficult they would be for a foreigner to learn particularly one whose mother tongue is English.

hammaslääkäri—dentist
kirjekuori—envelope
pyyheliina—towel
sanomalehti—newspaper
appelsiinimehu—orangeade
elokuvateatteri—cinema

Some other words are a bit easier!
pankki—bank

kahvi–coffee
muna–egg
kala–fish
kuuma–hot
joki–river

About 6 percent of Finns comprise the Swedish-speaking minority who mostly live in the coastal areas around Turku and Helsinki. Far more Finns than that are able to speak both languages fluently. Both Finnish and Swedish are official languages in the country and in places where there are large numbers of Swedish-speaking Finns street names, traffic signs, advertisements, and official government communications must be in both languages.

Lappish, another related language, is spoken by the Finnish Lapps in the north, although many Lapps also speak Finnish which they learn at school.

Finland's cultural heritage is long and varied which only makes

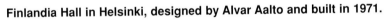

Finlandia Hall in Helsinki, designed by Alvar Aalto and built in 1971.

it possible to mention a few aspects here. The Finns are great readers and buying books is common among the whole population. Almost every home has a bookshelf with a good selection of books on it. Perhaps the long dark winter evenings are one reason for this.

In the past, in farming communities legends, stories, and epic poems were told again and again. These oral stories attracted scholars at the beginning of the 19th century and some of them began to record the old tales. Elias Lönnrot collected many of these stories and in 1835 he published *The Kalevala*, a book which made many Finns aware of the importance both of the Finnish language and of the ancient stories of the nation. *The Kalevala* tells the story of the myths of Finland and Karelia, and especially Väinämöinen (the god of music and poetry) and his brother Ilmarinen (the smith).

The Society of Finnish Literature has in its folklore archives in Helsinki the largest collection in the world of a nation's old stories consisting of 1,550,000 items.

A huge multi-volume work has also been collected and titled *The Ancient Poetry of the Finnish People*. Old Finnish folk songs have been gathered from different regions of the country, and these were often sung accompanied by the *kantele*, a Finnish instrument rather like a zither. The kantele has strings over a horizontal sounding board that are plucked by hand. Folk music, mainly from the country areas, has been collected in huge quantities with over 100,000 tunes now preserved.

Jean Sibelius (1865-1957) is Finland's most famous composer.

Ovenproof tableware in simple, attractive designs, typical of Finnish products. The mugs are made of brown glass.

His seven symphonies as well as the pieces called *Finlandia* and the *Karelia Suite* are played by orchestras all over the world. Sibelius's *Finlandia* was regarded as the unofficial national anthem as early as 1900 because the music seems to speak of the hopes of the whole nation. In Turku there is the Sibelius Museum which contains some of the composer's manuscripts and also musical instruments from many parts of the world. Many Finns enjoy music and there are a number of Finnish choirs, both professional and amateur. Most young people, however, as everywhere else in Europe, are interested in the latest pop songs on the charts.

Architecture is a popular subject of conversation in Finland and many Finns look at new buildings with interest and affection, yet also with a critical eye. The startling aspects of Finnish architec-

97

The sculpture of *The Three Smiths* in a busy street in Helsinki.

ture are its variety and surprises. Taivallahti Church consists of granite rock walls, with a roof of copper, glass, and wood. Alvar Aalto (1898-1976) was one of Finland's greatest architects, designing individual buildings such as Finlandia Hall (1971) in Helsinki, a concert and conference hall built of white marble, and also urban and regional plans and interior decoration, even designing furniture for his buildings.

There are sculptured figures everywhere in Finnish towns. Wäinö Aaltonen (1894-1966) used marble, ceramic, bronze, and granite and achieved startlingly different results.

Finland is a leading country in the field of industrial art by designing furniture, household utensils in stainless steel and wood, glassware, ceramics, and lamps, all of which are noted for simple, modern designs. The old Finnish woven rugs, once used on beds instead of blankets, are now made to hang on walls in Finnish homes. An organization known as The Friends of Finnish Handicrafts revive old handicrafts in textiles.

Helsinki has a Swedish theater and also the National Theater

98

which performs plays in Finnish. These are reminders that two groups of people have shaped the Finnish nation. In the past these were, first Swedish-speaking merchants, town dwellers, and farmers especially along the coast; and second, Finnish-speaking peasants in the inland forests. Slowly, over many years, these groups merged together to form the Finnish nation.

The Finnish Broadcasting Company began regular television transmissions in 1958 and every Finnish home now has a television set.

Public holidays in Finland are much the same as in other European countries but there are a few additional ones. On the evening before May Day (May lst), which was once a pagan festival to welcome the spring, students gather to enjoy themselves. In Helsinki they all meet at a fountain near the harbor and a student cap, white with a black peak, is placed on a statue's head. Balloons and toys are on sale from stalls in the Market Square. On May Day there are processions both of students and workers, for the latter have adopted May Day as Labor Day.

Midsummer is celebrated on a Saturday toward the end of June.

Children with the distinctive fair hair and blue eyes common among Finns.

The Finns believe in enjoying themselves. These are May Day celebrations in Helsinki. Students wear white caps with black peaks.

In the south thousands of bonfires—Finland's earthly suns—burn during the few hours in the middle of the night when the sun disappears below the horizon. National costumes from the different regions are sometimes worn on such special occasions. For women this includes a long skirt of bright colors, a long-sleeved blouse and a tightly-fitting tunic.

Independence Day is celebrated on December 6th when there are military parades in the capital, and candles are placed in the windows of houses.

One interesting Finnish custom is that everyone has a name day. (Every Christian name has a certain day in the calendar.) As this is regarded rather like a birthday, with gifts from family and friends, it is a very popular custom with Finnish children.

The welfare of the individual in Finland has improved greatly in recent years. The stress in social welfare schemes is on a better life for all. In such important aspects as sickness benefit, unem-

100

ployment benefit, public health, education, hospital treatment, and healthy conditions of work Finland is one of the leading countries in the world. Public expenditure has risen greatly as a result of such schemes. The government collects the money it needs to run the country from income tax on earnings and from a local tax payable to the community where the wage earner lives.

The number of babies born each year in Finland started to fall in the 1950s but this trend has now reversed. The fashion for having fewer children in each family was one cause of this decline in the birthrate; another was the emigration to Sweden (where wages are much higher) of many young married couples. The population growth has increased now to about 11 per 1,000 inhabitants each year. In 1995 Finland reported the lowest infant mortality in the world.

Finland has always been a country with a high proportion of women going out to work, and each area is required by law to provide day-care facilities for the children of working mothers. Finland is often at the forefront of progressive ideas and social changes and so it is hardly surprising that the idea of having a "mother's wage" was tried out. The plan was to find out how many women would decide to stay at home and look after their children if they received a wage for doing so paid from public money.

The Finns are a courageous, determined and contented nation to whom freedom means everything. They believe in working hard as well as in enjoying themselves. Winston Churchill said at the time of the Winter War: "Finland alone–in danger of death, superb, sublime Finland—shows what free men can do." The progress made by the Finnish nation in the 20th century illustrates what the Finns are capable of achieving in peace as well as war. They even have a word for their national characteristic—*sisu*—which means a mixture of stamina and courage, and a bit of stubbornness as well. It is this mixture in every Finn which has proved so useful in the past and, hopefully, will result in the building of an even better and more prosperous Finland for the future. As the novelist Johannes Linnankoski wrote of his native Finland:

This is a land with frost in the ground,
These are people with dreams.

GLOSSARY

berth   To bring a ship into harbor

brow-tine   Pointed branch of the reindeer antler

jääkiekko   Ice hockey

Kalevala   Book that contains the myths of Finland

kantele   Musical instrument of Finland

lichen   Type of moss which grows on hard surfaces

lorry   Automotive truck used for carrying freight

moor   To anchor a ship with cables

nomads   People who move from place to place usually in a defined territory looking for grazing land for flocks of animals

permafrost   Permanently frozen ground about two feet beneath the topsoil.

pesäpallo   Finnish form of baseball

ptarmigan   Type of bird in the grouse family that has feathered feet

pulkka   Small sled with one runner that is pulled by reindeer

reforestation	Renewing a forest by planting new trees as old ones are cut down
sauna	Finnish steam bath in which steam is provided by throwing water on hot stones
sisu	Character trait with a mixture of stamina and courage and a bit of stubbornness
tundra	Level treeless, boggy plain

INDEX